BIG YELLOW MACHINES

Written by Jean Eick
Illustrated by Michael Sellner

ABDO & Daughters

PUBLISHING

BIG YELLOW MACHINES

Published by ABDO & Daughters Publishing
4940 Viking Drive, Suite 622
Edina, Minnesota 55435 USA

Designed by Michael Sellner
Edited by Jackie Taylor
Production: James Tower Media • Design
Photo Credits: "Images © 1995 PhotoDisc, Inc."
and Caterpillar Image Lab

Printed in the United States of America

Library of Congress Cataloging-in-Publication Data
Eick, Jean, 1947-
    Bulldozers / written by Jean Eick; illustrated by Michael Sellner.
      p.   cm. – (Big yellow machines)
Summary: Introduces bulldozers, including their working parts, uses, fun
facts, and the inside of a bulldozer's cab.
    ISBN 1-56239-729-X
    1. Bulldozers – Juvenile literature.   [1. Bulldozers.]   I. Sellner, Michael, ill.
II. Title. III. Series.
TA725.E368   1996
629.225—dc20   96-4098
                                                                 CIP
                                                                 AC

Contents

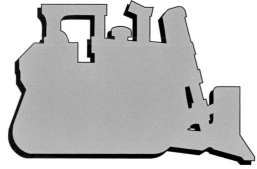

What is a Bulldozer?

Have you ever heard the roar of a bulldozer moving a gigantic pile of dirt or rocks? It is one of the most helpful machines for pushing heavy loads.

The "bulldozer" is actually the big curved blade in front. The machine is really just a big tractor, but most people call it a bulldozer.

Bulldozers don't have wheels. They move along on special tracks. A long time ago farmers used tractors like these to pull farm tools and deliver supplies.

Uses for Bulldozers

Bulldozers are very powerful machines. They can knock down trees and remove big rocks to make room for new roads, houses and buildings.

After a bulldozer has cleared an area, other big machines can go to work.

Huge bulldozers also work in coal mines. They use their giant blades to push the coal into piles and to load the trucks.

Special track shoes help the bulldozer move around without getting stuck in the coal.

Bulldozers can do many jobs. They can drag logs out of forests and bury garbage in landfills. Bulldozers can also clean up buildings destroyed by earthquakes and hurricanes.

During the Gulf War in Kuwait, bulldozers helped control huge oil fires.

With its special tools and crawler tracks, the bulldozer is a machine that can do a good job on almost any ground. No wonder it's used in so many different ways!

During World War II, soldiers used bulldozers as shields. They also used them to clean up areas that had been bombed.

Sometimes bulldozers were carried on rafts.

Today, bulldozers have many important jobs. They dig, carry and dump loads. They even help firefighters control forest fires. But when people think of bulldozers, the job they think of most is pushing huge walls of rocks and dirt.

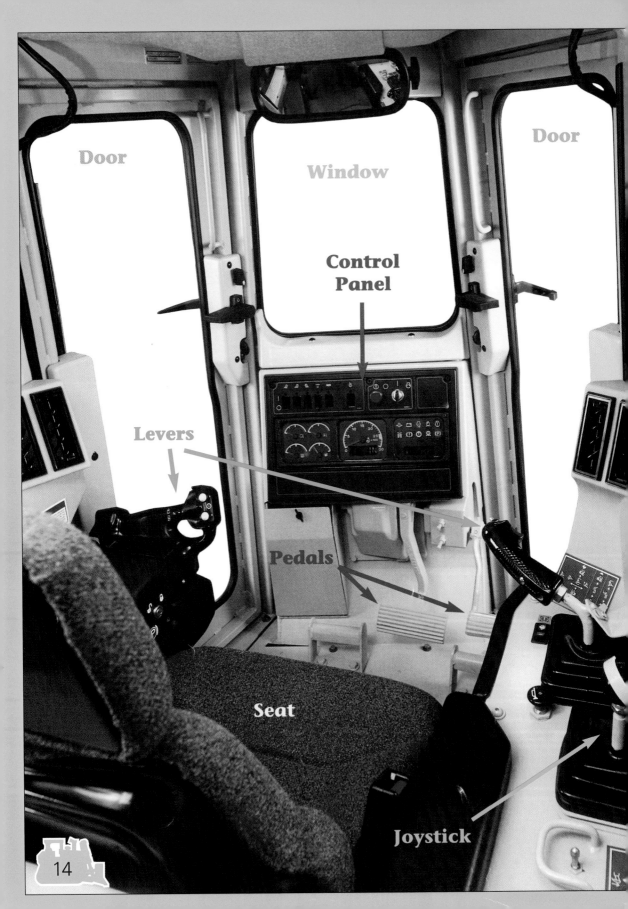

Door

Window

Door

Control
Panel

Levers

Pedals

Seat

Joystick

Inside the Cab

The cab is where the operator sits. Many cabs are open on the sides with a canopy over the top. When a job is very dirty or cold, the operator is protected inside a cab with windows and doors.

The operator uses pedals, levers and joysticks to steer the machine and move the blade. A control panel lets the operator know if everything is working.

The Parts of a **Bulldozer**

Canopy - A covering over the operator.

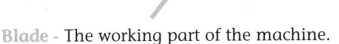

Blade - The working part of the machine.

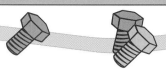

Cab - Includes the seat and controls for the operator.

Track roller frame - The frame around which the crawler tracks move.

Crawler tracks - Huge belts that take the place of tires.

Track shoes - The individual parts of the crawler tracks.

Engine - Where the power comes from to run the machine.

Tools of the Bulldozer

The bulldozer blade is the working part of the machine. It can do many, many things. Different blades have different jobs.

"A" Blade

An "A" blade can turn from side to side. It is used for clearing snow.

"PAT" Blade

A "PAT" blade can slant down to work in ditches. It's also the blade used on hills.

Trash Rack

A blade with a trash rack is used for working in landfills.

"S" Blade

The "S" blade pushes things straight ahead. It does a good job clearing land and lifting heavy loads.

A ripper is a tool added to the back of the tractor. Its strong, powerful tooth is like a giant claw that can rip apart concrete or pry out rocks buried deep in the ground.

One-Tooth Ripper

Three-Tooth Ripper

Where in the world can you find a
Bulldozer ?

Bulldozers help people do difficult jobs all around the world. They work in forests, deserts and mountains. Bulldozers have even worked inside volcanos!

North America

South America

Mountains

Forest

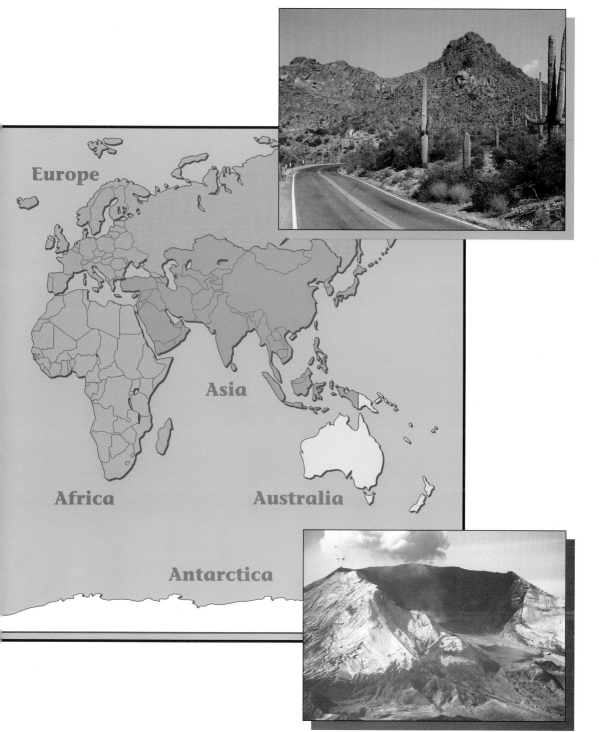

Desert

Europe

Asia

Africa

Australia

Antarctica

Volcano

Where in the world can you find a Bulldozer ?

Norway

In some areas of Norway, people have to drive a long way to visit each other or to shop. Bulldozers have helped build new roads to make traveling easier.

It snows a lot in Norway. Some people need bulldozers to bring them supplies during the winter.

Where in the world can you find a

Many areas of Mexico need more roads and houses. Bulldozers can help build them.

Some places also need more water. Bulldozers help build dams and canals to supply water all year long.

Where in the world can you find a Bulldozer ?

Russia

Ships carry bulldozers over to Russia. It takes them a long time to get there. Once, when gold miners needed more mining equipment, three giant bulldozers were sent over on the world's largest airplane.

Where in the world can you find a Bulldozer?

The bulldozer may be a slow mover, but it is a powerful worker. Its huge steel blade can knock over a brick wall or push back mud and dirt to help control the water of rivers raging over their banks. This rugged machine is a big help to people all over the world.

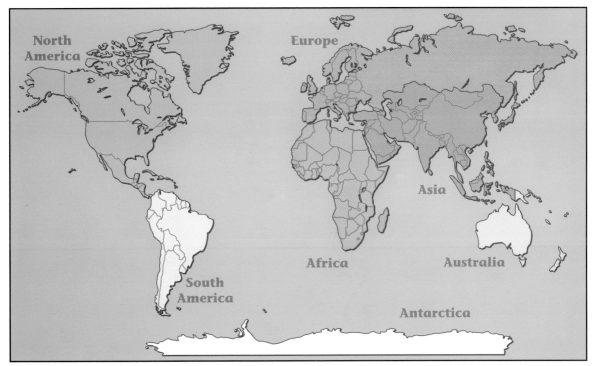

Special Facts About
Bulldozers

Bulldozers come in different sizes. Small bulldozers are powerful machines, but they seem tiny compared to the large bulldozers.

How Much Fuel Do Bulldozers Hold?

The small bulldozer holds 30.8 gallons (116.6 liters) of fuel. That could fill the tanks of two cars.

The large bulldozer could fill six schoolbus tanks with its 388 gallons (1,471 liters) of fuel.

How Tall Are Bulldozers?

The small bulldozer is as tall as an average size person (5 feet 6 inches/1.69 meters).

The large bulldozer is over twice as tall as an average size person (11 feet 6 inches/ 3.5 meters).

Small Bulldozer

🔩 How Much Do Bulldozers Weigh?

A small bulldozer weighs about as much as a full-grown elephant (15,518 pounds/7,039 kilograms).

The large bulldozer weighs much more than that.
It weighs 214,847 pounds (97,454 kilograms).
That's almost as much as a blue whale,
the heaviest animal in the world!

🔩 How Fast Can Bulldozers Go?

Bulldozers are tough, but they don't move very fast.
In fact, you could take a nice walk with a small bulldozer.
It goes 1.9 miles per hour (3.1 kilometers per hour).

The large bulldozer goes 7.2 miles per hour (11.6 kilometers per hour). You would have to run to keep up with that machine.

Large Bulldozer

Putting It All Together

It takes a lot of people to build one of these huge machines. Everyone has an important job making sure that all the parts are put together just right. Computers and robots check to make sure the bulldozer is ready for work before it leaves the factory.

This shiny new bulldozer is ready to go to work.

Look at that enormous engine!

Bulldozers move down the assembly line.

Words To Remember

 **Bulldozer** - A track-type tractor with a steel blade on the front of the machine.

 **Crawler tracks** - Huge steel belts that spin around a frame to move the machine.

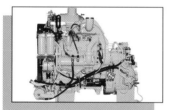 **Engine** - Where the power comes from to run the machine.

 **Track roller frame** - The frame around which the crawler tracks move.

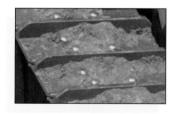 **Track shoes** - The individual parts of the crawler tracks.

 **Track-Type Tractor** - The earthmoving machine that is known as a bulldozer.

Index

32